The
Nevadaplano
A Brief Geology of the West

E. W. Eggleston

Manzanita Writers Press
San Andreas, CA

The Nevadaplano: A Brief Geology of the West

ISBN: 978-1-952314-17-9
Library of Congress Control Number: 2025909196

Published by:
 Manzanita Writers Press
 San Andreas, California
 manzanitawp@gmail.com

Cover Photo: The southern Sierra Nevada's Mount Whitney, at 14,505 feet, is the highest point in the continental United States. It is just 131 miles from Death Valley's Badwater Basin, which at 282 feet below sea level is the lowest point in North America.

Photographs, except as noted, are by E.W. Eggleston

Author: E.W. Eggleston
Contact him at buzz.calent@gmail.com

Dedication

To Karen, my soulmate and fellow traveler

The East Carson River meanders through the mountains just south of Markleeville, Alpine County.

Valley of Fire State Park, less than 50 miles northeast of Las Vegas, offers visitors grand views, petrified trees, and a tantalizing collection of indigenous rock art. Photo courtesy Nevada Division of State Parks.

Guide

Approaching Owens Valley from the east, with a skyline view of the highest peaks in the Sierra Nevada.

The West's Geology:
Looking Below the Surface

Years ago, I asked Google a few questions about the geology of the Sierra Nevada. One thing led to another and, well, it got out of hand. I read everything I could find, talked to experts across the West, wrote and shredded countless essays. I was disappointed and deeply frustrated with my inability to understand how the West was made. It didn't click. Could rocks really be that complicated?

Finally, I just put my research on a back shelf and went about the work of living the life of a retired country newspaper editor.

Months passed, then years. I took up other interests, traveled and enjoyed the good weather of California. That didn't last, though. Something kept nagging me, plaguing me with more questions about what had happened to the land. I needed to get to the bottom of it.

This is the story of what I learned when once again I went to look for answers. Like most things, it's best read in camp, in the shade of a big rock, with a cool beverage in hand.

One hundred twenty million years ago is a good place to start. You know the scene. You saw it in the movies. It was the Cretaceous Period: Dinosaurs ruled the land, the oceans teemed with strange creatures, the Earth was warm and moist, and what would become North America had long before drifted apart in big chunks from Pangea, the mother of continents.

In what would eventually be called the state of Nevada, some of Utah and their neighbors, a land heavily compressed by tectonic forces, the Nevadaplano was poised to become a very high plateau, although it would still take millions of years to reach its full prominence.

Nevadaplano? It's not a name I had heard before, but it turned out to be a missing signpost on my road to a better understanding of the West's complex geology.

I went to the source.

University of Arizona geologist Peter G. DeCelles at about 17,000 feet in the Kailas Range in the middle of the Tibetan Plateau. Photo courtesy Peter G. DeCelles.

The Plateaus

"The name is an homage to the Altiplano of western Bolivia, high in the central Andes," said Peter G. DeCelles, a geosciences professor at the University of Arizona who coined the word "Nevadaplano" in a 2004 research paper. He kindly answered my questions in an exchange of emails.

"(Bolivia's) Altiplano has an average regional elevation of about 12,000 feet—hence the 'alti'—and is mostly very flat—the 'plano,'" DeCelles continued. "The crust under the Altiplano is roughly twice normal thickness for continental crust, which explains why it is so high. The excess thickness results from the fact that western South America is caught in a vise between the eastward moving Nazca oceanic plate and the westward moving South American Plate.

"The situation in the western U.S. was very similar between about 100 million and 50 million years ago, with a very high—probably higher than 10,000 feet—'Nevadaplano' squeezed between westward moving North America and the eastward subducting Farallon Plate. I called it the Nevadaplano because it was mainly located in what is now the state of Nevada."

Long before there was a Nevadaplano, central Nevada was covered by a warm ocean. Wind-driven dust, sand and volcanic ash, along with runoff from nearby land, layered its depths. Slowly pushed up by the subducting Farallon Plate, dry land emerged and the ocean receded. Old reefs and a rich array of oceanic fossils are still found across nearly the entire state. *

* For more, see Appendix, "High Desert Treasures."

CANADA

Seattle
Olympia **Washington**

Portland
Salem
Oregon

Helena **Montana**
Fort Peck Lake

Idaho
Boise

Wyoming

Cheyenne

Great Basin
Reno
Sacramento • Carson City
San Francisco • Oakland • Stockton
San José
California
Fresno

Great Salt Lake
Salt Lake City
Provo

Nevada

Utah

Colorado
Denver
Colorado Springs

NORTH

Bakersfield
Las Vegas
Henderson

Los Angeles • Anaheim
Salton Sea
San Diego

PACIFIC

Arizona
Phoenix • Mesa

Tucson

Santa Fe
Albuquerque
New Mexico

El Paso

Texas

OCEAN

MEXICO

Credit: U.S. Fish and Wildlife Service map

Great Basin

Compressed by tectonics and by its own weight, a lot of old seabed became hard rock and the rock became mountains. The Nevadaplano rose up along with them. The high plain occupied the area of today's Great Basin, a 190,000-square-mile depression in Nevada's high desert. It's the northernmost part of the vast Basin and Range Province, which also includes Arizona, part of New Mexico, and extends deep into Mexico. When it rains on the Great Basin today, the rain stays there. No rivers flow out of it into any ocean. That wasn't the case when it was the Nevadaplano.

Some of the credit for what we know about how this happens begins with German meteorologist Alfred Wegener, who in 1912 came up with the idea of continental drift. He believed all continents had been part of a single landmass 300 million years ago and that they plow through the ocean floor much the same way as icebreakers plow through frozen seas. More than a half-century later his theory evolved into a widespread belief that continents do indeed move and the Earth is covered with tectonic plates, which also move.

These plates are enormous rock slabs generally found about 60 to 120 miles beneath our feet. The biggest is the Pacific Plate, which measures nearly 40 million square miles. Oceanic tectonic plates are primarily composed of relatively dense basaltic rock, and continental tectonic plates support lighter granitic rocks, the continents. They cover the entire Earth like the cracked shell of a boiled egg, a global jigsaw puzzle. About six miles thick where they form as crust at ocean ridges, and much thicker under continents, the plates collide with each other, dip below, or override one another, partly depending on their relative densities at the plate boundaries. They shape and reshape the world. Today there are 16 major and numerous minor tectonic plates around the world.

Mono Lake, a soda-saline body of water, formed after the explosive eruption that created the Long Valley Caldera and the Bishop Tuff more than 760,000 years ago. Pillars of tufa, formed above natural springs at the lake's bottom, are now exposed at the lake's surface.

Mono Lake is two to three times saltier than ocean water, making it a rich environment for brine and alkali flies. The brine flies thickly blanket the shoreline, attracting a wide variety of birds, including pelicans and gulls. The Kucadikadi, a band of Northern Paiute, harvested the pupae of the alkali flies, which when dried made for easily stored food.

Mono Lake

The plates move about, scientists believe, largely because of the Earth's internal heat and how it helps churn the mantle, a deep and molten part of the Earth's interior. That contributes to "push" where plates are generated and "pull" where plates are subducted.

It happens slowly. The Farallon slab passed under the continental plate traveling on average just a few centimeters—1.6 inches—per year. Yet it generated rock-melting heat, fueled by the release of highly compressed steam and other volatile materials squeezed from the subducting slab. This lowered the melting point in the overlying wedge of upper mantle to create magma, liquid or semi-liquid, which rises as it heats up.

The rising magma creates and fills cauldron-like chambers in the crust. When the magma eventually cools, the chambers harden and become a pluton; the name is a nod to Pluto, Greek god of the underworld. Where the Sierra Nevada is today, hundreds of magma-filled chambers became plutons. They range from less than a square mile in size to as much as 200 square miles and are equally impressive in their depth, sometimes measured in miles. As it cooled further, this collection of plutons became a batholith, the backbone of today's mountain range.

Eighty-seven million years ago, magma chambers and plutons remained buried in the Nevadaplano's western flank in a long, mostly north-south line called an arc. Most were still 2 to 10 miles below the Earth's surface. Magma roared upward in chimney-like columns from some of these chambers and fed stratovolcanoes, like those in today's Cascade Range. Just to the west of the volcanoes, the Pacific Ocean lapped primeval shores and rivers drained into it from the Nevadaplano, adding sediment to the seafloor.

DeCelles compared the volcanoes of the ancient Sierra Nevada to volcanoes in a different place and time. High stratovolcanoes of the western Andes sit in a similar position today relative to South America and its subducting Nazca Plate, he said.

A "metamorphic core complex" is a prize Nevada destination, the Ruby Mountains. The range in the northeastern part of the state invites visitors to experience dark skies, wilderness trails and grand vistas.

Forces of Nature

It's very easy to underestimate the magnitude of tectonic events and what they do over millions and tens of millions of years. A couple of centimeters a year adds up. They can push continents apart or bring them together. It's important to remember that geological time is difficult for humans to imagine. The rise and fall of the Nevadaplano is a case in point.

The Farallon Plate, called a slab once it subducted, traveled beneath the continent all the way to the Rocky Mountains, pushing them up, pushing up the Colorado Plateau, and pushing the land of Nevada upward as well.

The Nevadaplano was part of that. It was a province of Laramidia, a large island something like Japan, that was affected by the slab's passage below it. On the west, Laramidia was flanked by the Pacific Ocean. Its eastern border was near the Western Interior Seaway, a shallow waterway in today's Utah. The seaway separated Laramidia from another large island, Appalachia. As compression squeezed the land upward, the Western Interior Seaway retreated, creating a single landmass anchored by Laramidia and Appalachia like two bookends.

It was the birth of North America.

For millions of years the Nevadaplano stood on its western flank. At its height, the plateau might have exceeded much of today's Sierra Nevada in elevation. Deep valleys formed on its flanks. It was so tightly compressed and squeezed upward by tectonic forces that the place we now call Salt Lake City was 200 miles closer to Reno, about half the distance it is today.

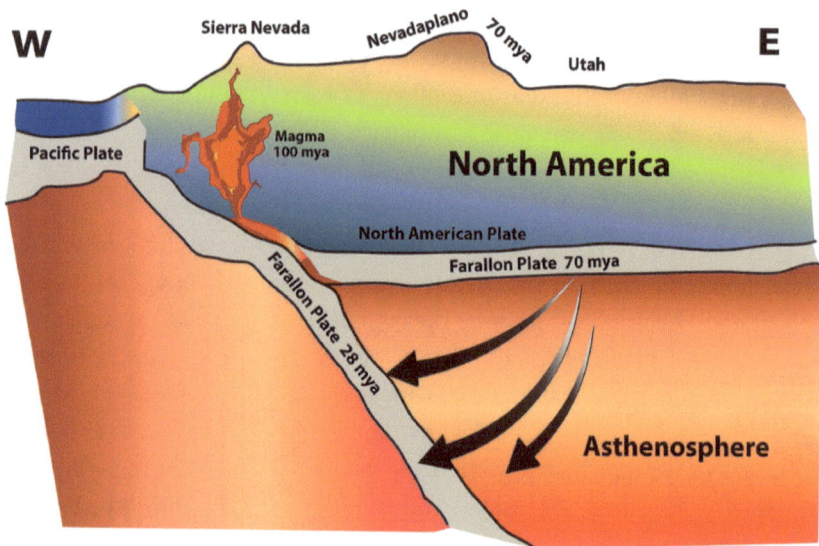

Credit: Debbie Wiegel, Design FX

Farallon Slab

When tectonic plates are subducted, they usually plunge quickly at a steep angle into the depths of the Earth, but almost from its start the subducting Farallon slab was different. It took a high road and moved at a shallow angle while hundreds of miles of continent passed over it. Then, about 50 million years ago, it began to drop into the mantle in a process geologists call "slab rollback." We would say it slumped, dragged down by immense weight, exposing the crust above it to the broiling heat of the upper mantle. The surface began to settle, and the Nevadaplano began to sink and spread out.

As the Farallon slab plunged more steeply into the Earth, super-heated molten rock from the earth's upper mantle rose to fill the growing void above the slab. Plumes of magma rose very close to the surface across central and southern Nevada, creating explosive eruptions. Some of those can still be seen today as calderas, large and crater-like, some many miles across. Deep volcanic tuff and broken rock, called ignimbrites, expelled by the eruptions, covers much of the region.

Hot Creek Geological Site near Mammoth Lakes is described by the U.S. Forest Service as "a place to marvel at geology in action. Boiling water bubbling up from the creek bed, fumaroles and periodic geyser eruptions at Hot Creek attest to the chamber of hot magma which lies about three miles below the surface of the earth in this area." Entering the water is prohibited because dangerous temperature changes can occur unexpectedly. Accessible hot springs are spread across Nevada, however, and guides to them are available online and where books are sold.

Birth of the San Andreas Fault

Beginning at least 36 million years ago, and continuing for about 18 million years, this "flare-up," as geologists call it, reshaped much of central and southern Nevada, and overlapped into neighboring states. During this time, just to the west, the last of the ancient volcanoes of the Sierra Nevada burned out, cooled, and slowly eroded.

The place where the Farallon Plate was born, its boundary with the Pacific Plate, is called the central East Pacific Rise. It was an ocean spreading center, so named because it is a place where magma rises out of the mantle and spreads out in opposite directions, creating new crust and pushing it outward. For eons it was a long, fuming mountainous crack at the bottom of the Pacific Ocean. About 28 million years ago, the rise was overridden by the North American Plate.

That event marked the start of the San Andreas Fault, a massive transform, strike-slip fault that now stretches nearly 800 miles, much of it along the California coast and beneath its offshore waters. The central Farallon Plate was completely subducted. Now the Pacific and North American plates are face to face there, and instead of one of them subducting, they slowly grind past each other. There are places in California, such as the Carrizo Plain, where you can put one foot on the Pacific Plate and the other foot on the North American Plate. Stand there long enough and you can feel the earth move. Well, no, most likely you can't feel it move, but you can imagine it.

Starting at about 5,300 feet elevation in the desert town of Baker, Nev., Great Basin National Park visitors drive the steep 12-mile-long Wheeler Peak Scenic Drive to a viewpoint about 3,000 feet below 13,063 foot Wheeler Peak, the park's highest summit.

One of the attractions of Great Basin National Park, near the Nevada-Utah border, are tours of Lehman Caves, itself a national monument with a rich history.

The Thinning Crust

The Pacific Plate pushes north, relatively, as geologists say. The North American Plate, moves south, again relatively. In truth, they both move generally northward, but the America Plate moves slower. That's what they mean when they say relatively.

By 15 million years ago, tectonic decompression had stretched Nevada in multiple directions and lowered its elevation. The rollback of the Farallon slab relieved tectonic pressure, causing what's called crustal extension. The Nevadaplano's elevation continued to drop, the high plain spread out, and then, abruptly, it was gone, replaced by the Great Basin, a depression in the land. Lakes formed on its surface as drainages were cut off, and parts of rivers that once flowed west toward the Pacific Ocean changed course and flowed east into the basin. The slumbering Sierra batholith beneath the high plain's western flank was exposed to sunlight for the first time, and spectacular new volcanic eruptions were triggered along the rising eastern flank of the Sierra Nevada.

The Earth's crust thinned throughout the entire Basin and Range. An "accordion-like" landscape of flat valleys and low mountain ranges, like falling dominoes, grew across the region. The thinning crust also created the abundance of natural hot springs we enjoy today and a significant geothermal industry. Today, 27 geothermal energy plants produce 8 percent of the state of Nevada's electricity, and scientists believe there's potential for it to generate much more.

A 19th-century mapmaker, Clarence Dutton, a colleague of Colorado River explorer John Wesley Powell, described the Basin and Range as a place where mountain ranges appeared like "an army of caterpillars" crawling north from Mexico. Today's satellite images support his ground-level observations.

A 2024 *Archaeology* magazine article describes the same area as a series of crustal blocks defined by "more than 560 long, parallel mountain ranges that reach heights of more than 14,500 feet." Of course, the highest of those blocks, the westernmost, would be the eastern escarpment of the Sierra Nevada. It forms the western wall of Walker Lane, a significant fault zone also called the Eastern California Shear Zone.

Backroads of the Great Basin often surprise travelers with rusty delights like these on display in Benton Hot Springs.

Walker Lane

Walker Lane, named after 19th-century pathfinder Joseph R. Walker, is a marching series of parallel earth fractures, a faultline. It generally follows the path of today's U.S. Highway 395 and it is crawling northward basically from the Gulf of California, the southern origin of the Salton Trough Fault. The northern end of that fault is closely tied to and believed to affect the San Andreas Fault at its origin near the Salton Sea in California's Mojave Desert.

The uplift of the Sierra Nevada was recent, about 5 million years ago, according to some scientists, and as much as 40 million years ago according to others. Recent studies, partly reconciling that difference, suggest the mountain range likely rose in widely separated pulses over the period. While the eastern side of the block was rising, the block's cooler and heavier western side sank deeply into California's Central Valley.*

Today's Great Basin averages less than 6,300 feet in elevation, probably less than half the height of the Nevadaplano's peaks, and Nevada is now the nation's fastest growing state, tectonically speaking. It adds about two acres per year thanks to ongoing crustal extension.

* For more, see Appendix, "Alabama Hills."

At 282 feet below sea level, Death Valley's Badwater Basin, an expansive salt flat, is the lowest point on the North American continent.

The Power of Erosion

More than tectonics has helped bring down its elevation, however. Throughout all time, gravity and rain have battled the Earth's efforts at mountain building. DeCelles' and others' research suggests hot, dry Nevadaplano summers helped draw monsoonal rains to areas on its flanks. Wind, rain and rivers erode mountains, steep slopes, deep canyons, and Nevadaplanos alike.

Finally, glaciers did some of the work of erosion, grinding away the landscape, sometimes sending torrents of mud and rock cascading downward. All of it, the rising mountains, volcanoes, wind, rain, and the movement and melting of ice, reshaped and sculptured the landscape.

The steep, sharp peaks of the Sierra Nevada, as well as the blunted and rounded ones, have all been shaped by erosion. It also is dramatically displayed in the "table mountains" that have formed at intervals along the western Sierra Nevada foothills. Towering above today's landscape, they were originally where successive rivers of lava flowed in pulses down toward California's Central Valley from volcanoes high above.

The river channels provided a mold for the cooling lava, which hardened with a "table"-like surface. The softer landscape alongside the channels then eroded over the ages, exposing solid lava walls sometimes a hundred feet high or more. Miners tunneled beneath them in the 19th century to look for placer gold in the ancient gravel riverbeds buried below the lava.

More staggering, though mostly hidden, evidence of erosion is the 3- to 6-mile-deep sediments of California's 450-mile-long Central Valley. All those high mountaintops and the bursts of ash, mud and rock from erupting volcanoes had to go somewhere.

Today, the Sierra Nevada presents a lofty fortress-like eastern wall, and spreading out from its base is the Great Basin.

U.S. Highway 50 crosses Nevada about mid-state and is nicknamed America's "Loneliest Road."

The Loneliest Road

Motorists on Highway 50, described by *Life* magazine in July 1986 as "The Loneliest Road in America," travel east or west across it. The road climbs a range of hills, drops, traverses a wide flat area, then climbs another range of hills only to repeat the exercise again and again. The pattern is disrupted when the route adjusts a bit to find a lower pass. Despite the "Loneliest" label, the road offers increasing numbers of visitors a geological adventure that also retraces the famed 1860-61 Pony Express route and gives them access to other wonders both historical and natural.

Throughout Nevada as well as most of the Basin and Range are volcanic fields, extensive lava flows, and cinder cones formed by the explosive eruption of basaltic rock, steam and gases. Some areas are deeply covered by volcanic ash and rubble.

The most notable eruption occurred fairly recently. Scientists think unlike most other calderas in the state, its origin is linked to the emerging Walker Lane. Iconic Mono Lake sits in a basin just north of a 20-by-11 mile caldera, one of the largest in the world, that erupted massively just 760,000 years ago and again, repeatedly, for at least a century afterward, blasting enormous amounts of ash, called the Bishop Tuff. It is a layer of compacted ash up to 4,900-feet deep that extends as far as Big Pine, 80 miles south.

Mammoth Mountain, though not considered an active volcano, has risen since that time. So did the Mono-Inyo Craters. These are a volcanic chain of relatively small craters, domes and lava flows that stretches 25 miles from the northwest shore of Mono Lake southward. It includes two volcanic islands in the lake and one cinder cone on its northwest shore. Eruptions are estimated to occur every 250 to 700 years along the chain. The most recent was an uplift of Paoha Island in Mono Lake about 250 years ago and was likely witnessed by humans, the ancestors of Paiute who still live in the area.*

* For more, see Appendix, "The Boom and Bust of a Mining Town."

Volcanic eruptions formed Negit, the dark island, and Paoha to its right in this photo, over the past 2,000 to 300 years as magma pushed up from below Mono Lake.

Panum Crater, a volcanic cone, formed just 600 years ago when rising magma and water combined explosively just south of Mono Lake. It's one of a chain of craters extending for 25 miles southward from the lake.

Eastern Sierra

Earthquakes strike regularly in the Eastern California Shear Zone, Walker Lane. Some of the quakes are big ones. On Independence Day, 2019, the city of Ridgecrest, Calif., a high desert town on Highway 395, experienced a 6.4 magnitude earthquake, followed the next day by a 7.1 magnitude quake with up to about 10 feet of right-lateral motion on the Paxton Ranch Fault.*

Five people were injured, but there were no deaths, and damage to buildings and infrastructure was estimated in the billions of dollars. There were 3,500 aftershocks in the next 10 days and many more since. Ridgecrest sits on a web of faults linked to Walker Lane.

A similar temblor had happened before, but with deadly consequences. In the very early morning of March 26, 1872, a major earthquake hit Owens Valley. It moved the ground in places as much as 40 feet horizontally. The epicenter, about 80 miles north of Ridgecrest, was Lone Pine, where 52 of the town's 59 dwellings collapsed. Losses totaled $250,000 (about $6.6 million today). More importantly, 27 people—about 10 percent of the town's extended population at the time—were killed, most of them in their beds. A small memorial sits on the north side of town to mark the spot where most of the earthquake's victims are buried.

Scientists, including James E. Faulds, professor with the Nevada Bureau of Mines and Geology at the University of Nevada, Reno, and formerly Nevada's state geologist, believe Walker Lane is slowly taking the place of the San Andreas Fault as the primary boundary between the Pacific and North American plates.

Faulds notes that as much as 25 percent of the motion between these plates has already been taken over by the Walker Lane fault zone and that the plate boundary has a history of stepping inland through time. He said the southern part of the fault shifted several million years ago from just west of Baja California into what is now the Gulf of California.

* For more, see Appendix, "A Noble Earthquake."

The Sherwin Range, including 12,241-foot Mount Morrison (center), provides a backdrop for meadows on the southeast edge of the Long Valley Caldera.

Sierra Nevada

The Sierra Nevada is not just its iconic granite peaks, but whole mountains and ranges of mountains, like the Ritter Range, made of erupted volcanic lava, steaming mud and ash, or the Sherwin Range, made of ancient, compacted sediments and now sitting as a "roof pendant" atop the Sierra batholith. Some of this likely includes leftovers from the Nevadaplano.

The geology of the Sierra Nevada has been described as "chaos," mainly because of this history and the abstract concretion of rocks that form its surface. A small army of geologists and their students have spent their careers mapping it and deciphering its geologic history.

DeCelles, drawing from his research, puts today's American West in a worldwide perspective.

"The Himalaya and Andes are currently active, in their constructive phases of development," DeCelles said, "whereas the western U.S. is about 50 million years ahead and well into its destructive phase." It is slowly disassembling in a process that will create a new world.

Peering into the future, Faulds has suggested that "the San Andreas Fault could be abandoned 5 million to 10 million years from now as the plate boundary takes another step inland to the Walker Lane in a belt extending from the Salton Sea in southern California, northward up the east side of the Sierra Nevada through western Nevada, and even into southwestern part of Oregon."

In an even bigger picture, today's Pacific Northwest geology is somewhat like California's geology of more than 45 million ago, but with a significant difference. The Juan de Fuca and Gorda plates, remnants of the Farallon Plate, abut far Northern California, Oregon, Wasington, and southern British Columbia and are subducting beneath the North American Plate. Unlike their parent plate, however, their descent is very steep and they will be completely subducted in about 17 million years.

The backroads of the Great Basin routinely offer up geological surprises such as this fractured monolith on a roadside near Benton Hot Springs.

The Lesson

California, and indeed the world, continues to move. Some geologists believe that millions of years from now a California "microplate," perhaps together with Baja California, will be pushed north toward Alaska, creating beachfront land in the deserts of Arizona and Nevada.

So, after all of this, what have I learned in my journey to understand the geology of the West?

Well, for one thing, geologists, especially those who have put up with my interminable questions, are as patient as the mountains they study. For another, in the geological world, time is much longer and things are often vastly bigger than we mere humans can imagine.

I also learned that what you see on the surface, beautiful as it is, was mostly created by volcanoes or at least the stuff that makes volcanoes, and just as often by sediments that settled in thick layers in shallow seas. It is rock, mud, sand, or ash, pummeled and shaped by water and wind, sometimes compressed, heated and layered and twisted into fantastic shapes, and often reshaped repeatedly by the whims of nature or eroded and reshaped some more.

Deep beneath what we can see on the surface are the titanic, tectonic forces that do the heavy work of building continents.

Information about the West's geology is still rushing in. It often changes what we thought we knew for certain, and no one knows all there is to know about it. The Nevadaplano is just a small piece in the great puzzle that is the geology of the western United States. Understanding its place in the puzzle, however, makes it easier to see the bigger picture of how this landscape formed, and perhaps it gives us a glimpse of how this place might appear in the distant future.

And yes, finally, there's one more important thing I've learned about geology: Rocks really are that complicated!

Appendix

At 7,000-feet on the western slopes of the Shoshone Mountains, Berlin-Ichthyosaur State Park was established in 1957 to protect North America's largest collection of ichthyosaur fossils.

Visitors to the Berlin-Ichthyosaur State Park can experience both a major paleontology site and one of Nevada's early deep mines.

A rusting Model T on display at the Berlin Mine.

High Desert Treasures

Berlin-Ichthyosaur State Park, mid-state, about 50 miles south of Highway 50, reveals two remarkable stories of Nevada natural history: its connections to an ancient sea and to humanity's unrelenting quest for prized minerals.

This remote park is home to the largest known concentration of Ichthyosaur remains in the world. About 40 specimens of the large, ancient marine reptile, which became extinct about 90 million years ago, have been found there since the first discovery in 1928, and many are now displayed at the park's Fossil House.

The fossil discovery site is next to the Berlin Mine and a townsite of the same name. Silver was first discovered there in 1863, but serious mining did not get underway until decades later.

The landscape of Nevada and eastern California overlies a treasure of mineral wealth created by nature's geologic furnaces. Precious metals are sometimes found in veins, mostly quartz, that rose as a dense, hot fluid in cracks through layers of ancient rock, and then it cooled. Sometimes they are found as invisibly small particles diffused throughout dense rock formed in the anvil of compression. In either case, the valuable metals are highly sought after by men and women who have mined the state commercially since the latter half of the 19th century.

The buildings at Berlin, including the mill, are kept in "arrested decay," a sort of curated ghost town where visitors are free to meander. The mine, established in 1886-1887, successfully produced $849,000 worth of gold and silver from three miles of underground tunnels during its lifetime. However, A 1907 worker strike crippled operations and by 1914 all mining had stopped there. Elsewhere in the state, though, the boom continues.

The federal Bureau of Land Management records 1,241,156 mining claims in Nevada, 231,922 of them active, far more than any other state. Mining is a critical part of the Silver State's economy, second only to tourism and gambling. It directly employs at least 15,000 people and many thousands more earn their livelihood in related fields.

The Ancient Bristlecone Pine Forest in the White Mountains east of the Owens Valley is home to some of the world's oldest living trees. A core taken from Methusalah in 1957 dated the tree back 4,789 years.

John Muir,
Photo by Bailey Millard, circa 1893.
Photo courtesy U.S. National Park Service

A Noble Earthquake

There were no seismometers when the 1872 Owens Valley earthquake struck, but it has been estimated at between 7.4 and 7.9 magnitude. It was felt as far south as Tijuana and as far north as Oregon.

John Muir, 115 miles away in Yosemite Valley, awoke and dashed out of his cabin to experience the quake from a perch in a pine tree. He called it "a noble earthquake" and documented the experience in his journal and later in his 1901 book *Our National Parks*.

"The Eagle Rock, a short distance up the valley, had given way, and I saw it falling in thousands of the great boulders I had been studying so long, pouring to the valley floor in a free curve luminous from friction, making a terribly sublime and beautiful spectacle—an arc of fire fifteen hundred feet span, as true in form and as steady as a rainbow, in the midst of the stupendous roaring rock-storm. The sound was inconceivably deep and broad and earnest, as if the whole earth, like a living creature, had at last found a voice and were calling to her sister planets. It seemed to me that if all the thunder I ever heard were condensed into one roar it would not equal this rock roar at the birth of a mountain talus."

Bodie is a ghost town in the Bodie Hills

The Boom and Bust of a Mining Town

Mark Twain described the change of season in the mining town of Bodie as ". . . the breakup of one winter and the beginning of another," where weather killed as many men as bullets.

Designated a California state historic park in 1962, Bodie has a notorious history.

It was founded in 1859 after W.S. Body (or Bodey) found small amounts of gold in the rolling hills north of Mono Lake. The discovery of a rich vein of gold in 1875 attracted fortune hunters to the boomtown and created a lawless Wild West scene where gambling saloons and brothels prospered alongside general stores, blacksmiths, and boarding houses.

But the richest ore soon played out and the population fell dramatically. Two disastrous fires in 1892 and 1932 also took a toll.

"Mining officially ceased in Bodie in 1942, the final nail in the coffin for Bodie's township," a state park's department history says. Fewer than 10 percent of the 2,000 structures remained.

"Only a small part of the town survives. Interiors remain as they were left and stocked with goods. Designated as a National Historic Site and a State Historic Park in 1962, the remains of Bodie are being preserved in a state of 'arrested decay.' Today this once thriving mining camp is visited by tourists who continue to travel to Bodie from all over the world, just as they did back in Bodie's boom years," the state history says.

Visitors to Bodie stroll through a ghost town of abandoned buildings, including glimpses of life in the 19th and early 20th centuries when comforts we take for granted were non-existent.

The entry road to Red Rock Canyon State Park on the southern stretch of Highway 395, teases a preview of the stunning rock forms to be found in the park.

The Alabama Hills National Scenic Area.

Alabama Hills

If your first impression of the Alabama Hills is a sense of familiarity, don't be surprised. Hollywood has been making films there since 1920, many of them the blockbusters of their day.

The Alabama Hills National Scenic Area outside of Lone Pine, Calif., has provided the set for more than 700 movie and television productions since Roscoe "Fatty" Arbuckle mounted up to film the 1920 wester *The Round Up*. The classic *Gunga Din* was made there in 1939. More recently, it has been the backdrop for scenes in *Tremors, Star Wars, Iron Man, Gladiator, The Lone Ranger, and Django Unchained*, to name a few. A museum dedicated to the history of filmmaking in the Alabama Hills is a popular stop in Lone Pine.

Geologically, the Alabama Hills are made of the same granitic rock that forms the massive mountains just to their west, the Sierra Nevada. They've just experienced a difference in chemical and wind erosion, with glaciers carving the high peaks of the Sierra while wind, rain and temperature molded and sculpted the much lower Alabama Hills.

Southern sympathizers who were prospecting in the area in the 1860s named the hills in honor of the CSS Alabama, a confederate warship. The name has stuck despite attempts by some to change it.

More recently, efforts have been made to better regulate visits to the area, especially camping, and to preserve its historical, geological and botanical resources for future generations to enjoy.

A small area braced against the southern Sierra Nevada, the Alabama Hills, attracts thousands of visitors every year.

Look But Don't Touch

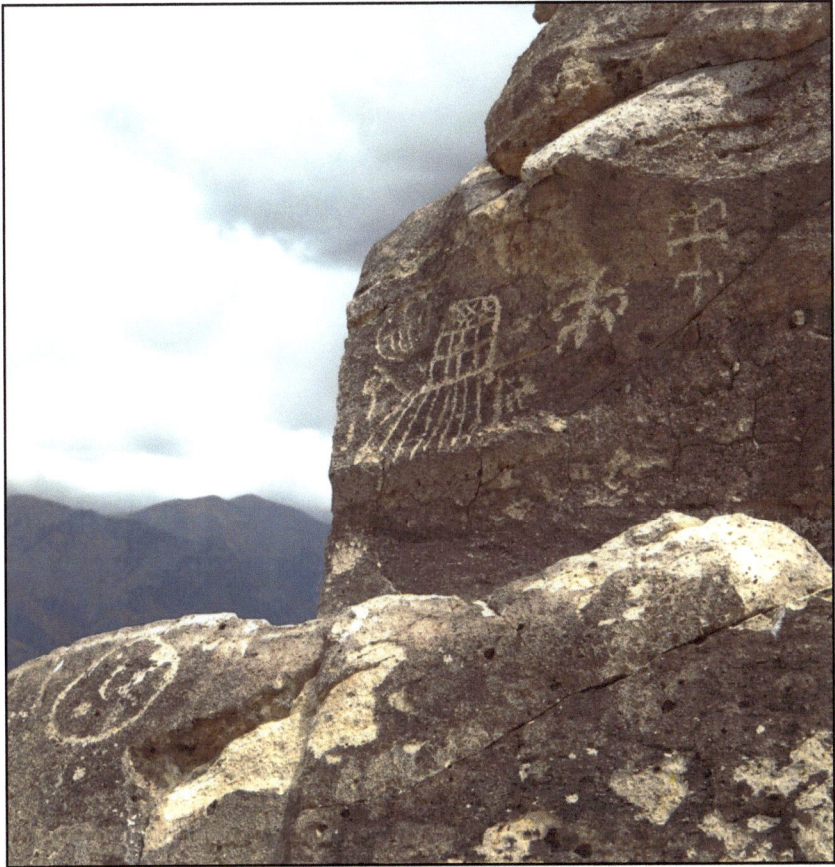

Petroglyphs and other ancient rock art, some thousands of years old, are found throughout Nevada. In some places—like Grimes Point/Hidden Cave Archaeological Site near Fallon—viewing them is encouraged with walkways and signage. Visitors are asked not to touch rock art anywhere, ever, however, and to stay on trails. Even a small amount of the oils from our hands can darken petroglyphs and damage pictographs, in time making them impossible to see. Please look, but don't touch. Federal and state laws can lead to fines, probation or imprisonment for damaging rock art or for removing artifacts.

E.W. "Buzz" Eggleston

About the Author

E.W. (Buzz) Eggleston is a retired newspaper editor. He resides in San Andreas, Calif., is married (Karen), has a daughter and two grown grandchildren. His career led him to work at five newspapers over 40 years, beginning with the *Palo Alto Times*, which in 1980 assigned him to cover the deadly eruption of Mount St. Helens in Washington state. That initially inspired his interest in geology. Later, he edited community newspapers in the Mother Lode, the heart of California's gold country in the Sierra Nevada foothills, which increased his interest in earth sciences, including wildlife and plant diversity. Today he enjoys travel, camping and photography. Contact him at buzz.calent@gmail.com.

The writer extends his deep appreciation to Professor Peter G. DeCelles of the University of Arizona, to Professor James E. Faulds of the University of Nevada, Reno, to Cathy Busby, emeritus professor at the University of California, Davis, to George Bergantz, emeritus professor at the University of Washington, and to other geologists and their students who have worked so hard to make sense of geology and shared their findings with us. Any errors are the sole responsibility of the author.

Notes

Notes

Perhaps the most photographed cabin in the Sierra Nevada, this simple dwelling rests in a meadow alongside Highway 88 (Carson Pass) on the western fringe of Hope Valley. The view changes with every season.

www.ingramcontent.com/pod-product-compliance
Lightning Source LLC
Chambersburg PA
CBHW040938030426
42335CB00001B/29